I'm a THREE Times Tabler!

Bill Gillham and Mark Burgess

A Magnet Book

Where are all these cats going?

Why are they all in threes?

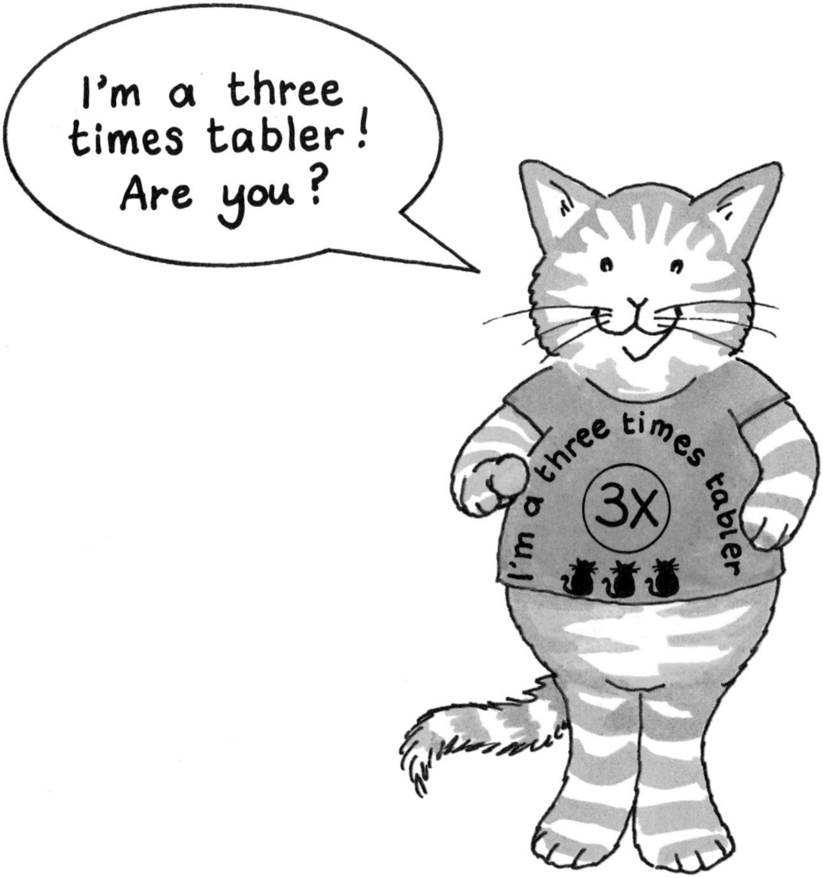
I'm a three times tabler! Are you?
I'm a three times tabler
3X

$$1 \times 3 = 3$$

2 × 3 = 6

$3 \times 3 = 9$

$$4 \times 3 = 12$$

$$5 \times 3 = 15$$

$$6 \times 3 = 18$$

six threes are eighteen

$7 \times 3 = 21$

$$8 \times 3 = 24$$

$$9 \times 3 = 27$$

$$10 \times 3 = 30$$

11 × 3 = 33

$$12 \times 3 = 36$$

$1 \times 3 = 3$

$2 \times 3 = 6$

$3 \times 3 = 9$

$4 \times 3 = 12$

$5 \times 3 = 15$

$6 \times 3 = 18$

$7 \times 3 = 21$

$8 \times 3 = 24$

$9 \times 3 = 27$

$10 \times 3 = 30$

$11 \times 3 = 33$

$12 \times 3 = 36$

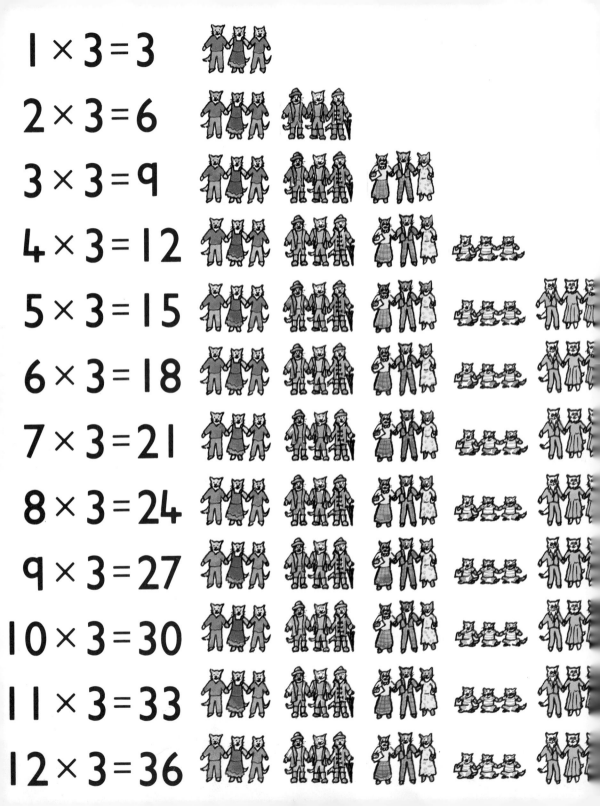

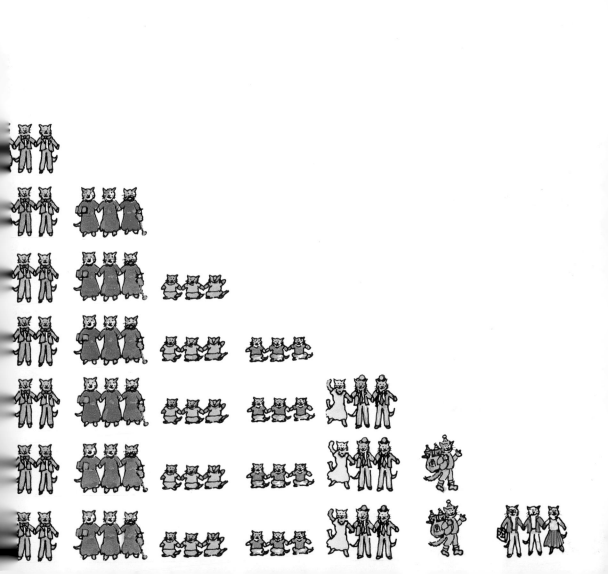

Activities

TRAFFIC LIGHTS

On a long strip of paper (sellotape sheets together
if necessary) draw twelve traffic lights in a row.
The child can colour in the lights – red, orange
and green – and then recite the table along the line.

A slightly more expensive alternative is to stick
fruit gums on the traffic lights.
If you want to draw a longer 'road'
then a toy car can be made to stop at the lights
as the child counts in threes: 3 – 6 – 9 etc.

BEAD THREADING

Start the child off with threading the first three beads
(all the same colour), showing how you choose
a different colour for each 'set' of three.
Thread up to **36** then ask him to recite the table
up the lace.

Explain that when you say the three times table
you are *counting* in threes: do that next.

THREE BEARS

Make twelve of each
of the three bears
by tracing over
the figures below:

Mix them up and ask the child to sort into
'families'. This is best done by sticking them
on to a long sheet of paper, and children will
enjoy colouring them in or writing *daddy*,
mummy and *baby* underneath.

Encourage the child to say the three times table
along the line and then ask,
'How many bears altogether?' and,
'How many families?'

NUMBER LINE

On a long, narrow strip of paper (cut A4 sheets
lengthways and sellotape together) write
in *large* numerals the numbers 1 to 36
in black felt-tip pen but leaving gaps for
3, 6, 9, 12, 15, 18, 21, 24, 27, 30, 33 and 36.

Ask the child to fill in the missing numbers
in red felt-tip and then to say the red numbers
out loud.

If the child has any difficulty in writing numerals
you can put them in lightly in pencil so that
they can be copied over in red.

Children need to know their tables because:
— simple multiplication, *which you can do in your head,*
is a skill of practical use in everyday life;
— the number patterns and groupings that occur
in tables help them to understand more advanced
mathematical concepts like *sets,* number *series*
and *progressions.*

The Times Table Books teach these ideas in a clear
and enjoyable fashion and show vividly what happens
when you multiply.

*Dr Bill Gillham is senior lecturer in the
Department of Psychology at Strathclyde University.*

First published in Great Britain in 1987
as a Magnet original
by Methuen Children's Books Ltd
11 New Fetter Lane, London EC4P 4EE
Text copyright © 1987 Bill Gillham
Illustrations copyright © 1987 Mark Burgess
Printed in Great Britain

ISBN 0 416 00212 9